法 兵 子 孫

Sun Tzŭ

ON THE

Art of War

edited by
MAJOR THOMAS R. PHILLIPS

Martino Fine Books
Eastford, CT
2018

Martino Fine Books
P.O. Box 913,
Eastford, CT 06242 USA

ISBN 978-1-68422-254-4

Copyright 2018
Martino Fine Books

All rights reserved. No new contribution to this publication may be reproduced, stored in a retrieval system, or transmitted, in any form or by any means, electronic, mechanical, photocopying, recording, or otherwise, without the prior permission of the Publisher.

Cover Design Tiziana Matarazzo

Printed in the United States of America On 100% Acid-Free Paper

孫 子 兵 法

SUN TZŬ

ON THE

ART OF WAR

edited by
MAJOR THOMAS R. PHILLIPS

LONDON
John Lane the Bodley Head

First published in England 1943
Copyright in U.S.A.

This book is produced in complete conformity with the Authorized Economy Standards.

Printed and made in Great Britain by
STEPHEN AUSTIN AND SONS, LTD, HERTFORD
for JOHN LANE THE BODLEY HEAD LIMITED
8 Bury Place, London WC1

SUN TZU
The Art of War
Introduction

WRITTEN about 500 B.C., THE ART OF WAR, by Sun Tzu, is the oldest military treatise in the world. Highly compressed, it is devoted to principles and still retains its original value. To the military student able to adapt its principles to modern warfare it is, even two thousand four hundred years after its preparation, a valuable guide for the conduct of war. Although the chariot has gone and weapons have changed, this ancient master holds his own, since he deals with fundamentals, with the influence of politics and human nature on military operations. He shows in a striking way how unchanging these principles are.

The tactics and strategy of Sun Tzu place high value on manœuvre. He advises the avoidance of battle unless all considerations are favourable. Victory is won by indirect methods. The holding force and the enveloping force are the direct and indirect elements of the army. 'In all fighting the direct method may be used for joining battle, but indirect methods will be needed to secure victory.' And he adds that indirect tactics are as inexhaustible as Heaven and Earth, as unending as the flow of rivers and streams. 'All men,' he repeats, 'can see the tactics whereby I conquer, but what none can see is the strategy out of which victory is evolved.' The victorious strategist seeks battle after the victory has been won, while he who is destined to defeat first fights and seeks victory afterwards.

'All warfare is based on deception.' When able to attack, the general must seem unable; when using his forces he should appear inactive. Tactical dispositions should mask strength with weakness. And conversely, the skilful opponent should be judged, not by appearances, but by the more precise information obtained from spies. The five kinds of spies are listed and their use is treated in detail. Inward spies are officials of the enemy. Sun Tzu could well understand how the German aviators were able to bomb Polish headquarters every time it was moved in September, 1939. Converted spies are enemy spies who have been bought off.

Estimation and calculation are given a wholly modern place in gaining victory. Ponder and deliberate before making a move. Sun Tzu claims he can forecast victory or defeat by the seven items of his estimate. His large attention to terrain is also in accord with modern practice. 'The natural formation of the country is the soldier's best ally.' Nor does he forget the importance of discipline and the need for gaining the loyalty of his soldiers. 'Regard your soldiers as your children, and they will follow you into the deepest valleys; look on them as your own beloved sons, and they will stand by you unto death.' Men must be gradually led to discipline. 'If soldiers are punished before they have grown attached to you they will not prove submissive. If, when the soldiers have become attached to you, punishments are not enforced, they (the soldiers) will be useless.'

No page of Sun Tzu's book can be read without finding the distilled wisdom of a great soldier written with the aphoristic distinctness of Chinese literature. THE ART OF WAR is the greatest military classic of the Chinese. It has had innumerable commentators and has been plagiarized throughout its existence. It is held in even greater reverence in Japan than in China, where the low estate of the soldier prevented the literary recognition it warrants.

The text of THE ART OF WAR has been transcribed from the translation by Lionel Giles, M.A., Assistant in the Department of Oriental Books and Manuscripts in the British Museum. It was first published by Luzac and Co., London, in 1910, and is still available (10*s*. 6*d*. net). The critical notes of the translator, which comprise the larger portion of his book, are here omitted. Other translations in English and French are lacking both in accuracy and the crystalline language which Dr. Giles has given his work. Grateful acknowledgment is made to Dr. Giles and Luzac and Co. for their generous permission to include his translation in this collection.

The Art of War

I

LAYING PLANS. Sun Tzu said: The art of war is of vital importance to the state.

It is a matter of life and death, a road either to safety or to ruin. Hence it is a subject of inquiry which can on no account be neglected.

The art of war is governed by five constant factors, to be taken into account in one's deliberations, when seeking to determine the conditions obtaining in the field.

These are: (1) The Moral Law; (2) Heaven; (3) Earth; (4) The Commander; (5) Method and Discipline.

The Moral Law causes the people to be in complete accord with their ruler, so that they will follow him regardless of their lives, undismayed by any danger.

Heaven signifies night and day, cold and heat, times and seasons.

Earth comprises distances, great and small; danger and security; open ground and narrow passes; the chances of life and death.

The Commander stands for the virtues of wisdom, sincerity, benevolence, courage, and strictness.

By *Method and Discipline* are to be understood the marshalling of the army in its proper subdivisions, the gradations of rank among the officers, the maintenance of roads by which supplies may reach the army, and the control of military expenditure.

These five heads should be familiar to every general; he who knows them will be victorious; he who knows them not will fail.

Therefore, in your deliberations, when seeking to determine the military conditions, let them be made the basis of a comparison, in this wise:

(1) Which of the two sovereigns is imbued with the Moral law?
(2) Which of the two generals has most ability?
(3) With whom lie the advantages derived from Heaven and Earth?
(4) On which side is discipline most rigorously enforced?
(5) Which army is the stronger?
(6) On which side are officers and men most highly trained?
(7) In which army is there the greater constancy both in reward and punishment?

By means of these seven considerations I can forecast victory or defeat.

The general that harkens to my counsel and acts upon it will conquer—let such a one be retained in command! The general that harkens not to my counsel nor acts upon it will suffer defeat—let such a one be dismissed! While heeding the profit of my counsel, avail yourself also of any helpful circumstances over and beyond the ordinary rules. According as circumstances are favourable, one should modify one's plans.

All warfare is based on deception. Hence, when able to attack, we must seem unable; when using our forces, we must seem inactive; when we are near, we must make the enemy believe that we are away; when far away, we must make him believe we are near. Hold out baits to entice the enemy. Feign disorder and crush him.

If he is secure at all points, be prepared for him. If he is superior in strength, evade him. If your opponent is of choleric temper, seek to irritate him. Pretend to be weak that he may grow arrogant.

If he is taking his ease, give him no rest. If his forces are united, separate them. Attack him where he is unprepared, appear where you are not expected. These military devices, leading to victory, must not be divulged beforehand.

Now the general who wins a battle makes many calculations in his temple ere the battle is fought. The general who loses a battle makes but few calculations beforehand. Thus do many calculations lead to victory, and few calculations to defeat: How much more no calculation at all! It is by attention to this point that I can see who is likely to win or lose.

2

WAGING WAR. Sun Tzu said: In the operations of war, where there are in the field a thousand swift chariots, as many heavy chariots, and a hundred thousand mail-clad soldiers, with provisions enough to carry them a thousand *li*,[1] the expenditure at home and at the front, including entertainment of guests, small items such as glue and paint, and sums spent on chariots and armour, will reach the total of a thousand ounces of silver per day. Such is the cost of raising an army of 100,000 men.

When you engage in actual fighting, if victory is long in coming, the men's weapons will grow dull and their ardour will be damped. If you lay siege to a town you will exhaust your strength. Again, if the campaign is protracted, the resources of the State will not be equal to the strain.

Now, when your weapons are dulled, your ardour damped, your strength exhausted, and your treasure spent, other chieftains will spring up to take advantage of your extremity. Then no man, however wise, will be able to avert the consequences that must ensue.

Thus, though we have heard of stupid haste in war, cleverness has never been associated with long delays. There is no instance of a country having been benefited by prolonged warfare.

It is only one who is thoroughly acquainted with the evils of war that can thoroughly understand the profitable way of carrying it on. The skilful soldier does not raise a second levy, neither are his supply-wagons loaded more than twice. Bring war material with you from home, but forage on the enemy. Thus the army will have food enough for its needs.

Poverty of the state exchequer causes an army to be maintained by contributions from a distance. Contributing to maintain an army at a distance causes people to be impoverished.

On the other hand, the proximity of an army causes prices to go up; and high prices cause the people's substance to be drained away.

When their substance is drained away the peasantry will be afflicted by heavy exactions.

[1] 2·78 modern *li* equal one mile.

With this loss of subsistence and exhaustion of strength the homes of the people will be stripped bare and three-tenths of their incomes will be dissipated; while Government expenses for broken chariots, worn-out horses, breast-plates and helmets, bows and arrows, spears and shields, protective mantlets, draught-oxen and heavy wagons, will amount to four-tenths of its total revenue.

Hence a wise general makes a point of foraging on the enemy. One cartload of the enemy's provisions is equivalent to twenty of one's own, and likewise a single picul of his provender is equivalent to twenty from one's own store.

Now in order to kill the enemy our men must be roused to anger; that there may be advantage from defeating the enemy they must have their rewards.

Therefore in chariot fighting, when ten or more chariots have been taken, those should be rewarded who took the first. Our own flags should be substituted for those of the enemy, and the chariots mingled and used in conjunction with ours. The captured soldiers should be kindly treated and kept. This is called using the conquered foe to augment one's own strength.

In war, then, let your great object be victory, not lengthy campaigns.

Thus it may be known that the leader of armies is the arbiter of the people's fate, the man on whom depends whether the nation shall be in peace or peril.

3

ATTACK BY STRATAGEM. Sun Tzu said: In the practical art of war the best thing of all is to take the enemy's country whole and intact; to shatter and destroy it is not so good. So, too, it is better to capture an army entire than to destroy it, to capture a regiment, a detachment, or a company entire than to destroy them.

Hence to fight and conquer in all your battles is not supreme excellence; supreme excellence consists in breaking the enemy's resistance without fighting.

Thus the highest form of generalship is to baulk the enemy's plans; the next best is to prevent the junction of the enemy's forces; the next in order is to attack the enemy's army in the field; and the worst policy of all is to besiege walled cities.

The rule is not to besiege walled cities if it can possibly be avoided. The preparation of mantlets, movable shelters, and various implements of war, will take up three whole months; and the piling up of mounds over against the walls will take three months more.

The general, unable to control his irritation, will launch his men to the assault like swarming ants, with the result that one-third of his men are slain, while the town remains untaken. Such are the disastrous effects of a siege.

Therefore the skilful leader subdues the enemy's troops without any fighting; he captures their cities without laying siege to them; he overthrows their kingdom without lengthy operations in the field.

With his forces intact he will dispute the mastery of the Empire, and thus, without losing a man, his triumph will be complete. This is the method of attacking by stratagem.

It is the rule in war, if our forces are ten to the enemy's one, to surround him; if five to one, to attack him, if twice as numerous, to divide our army into two.

If equally matched, we can offer battle; if slightly inferior in numbers, we can avoid the enemy; if quite unequal in every way, we can flee from him. Hence, though an obstinate fight may be made by a small force, in the end it must be captured by the larger force.

Now the general is the bulwark of the state: if the bulwark is complete at all points, the state will be strong; if the bulwark is defective, the state will be weak.

There are three ways in which a ruler can bring misfortune upon his army:

(1) By commanding the army to advance or to retreat, being ignorant of the fact that it cannot obey. This is called hobbling the army.

(2) By attempting to govern an army in the same way as he administers a kingdom, being ignorant of the conditions which obtain in an army. This causes restlessness in the soldiers' minds.

(3) By employing the officers of his army without discrimination, through ignorance of the military principle of adaptation to circumstances. This shakes the confidence of the soldiers.

But when the army is restless and distrustful, trouble is sure to come from other feudal princes. This is simply bringing anarchy into the army, and flinging victory away.

Thus we may know that there are five essentials for victory: (1) He will win who knows when to fight and when not to fight. (2) He will win who knows how to handle both superior and inferior forces. (3) He will win whose army is animated by the same spirit throughout all ranks. (4) He will win who, prepared himself, waits to take the enemy unprepared. (5) He will win who has military capacity and is not interfered with by the sovereign. Victory lies in the knowledge of those five points.

Hence the saying: If you know the enemy and know yourself you need not fear the result of a hundred battles. If you know yourself but not the enemy, for every victory gained you will also suffer a defeat. If you know neither the enemy nor yourself you will succumb in every battle.

4

TACTICAL DISPOSITIONS. Sun Tzu said: The good fighters of old first put themselves beyond the possibility of defeat, and then waited for an opportunity of defeating the enemy.

To secure ourselves against defeat lies in our own hands, but the opportunity of defeating the enemy is provided by the enemy himself.

Thus the good fighter is able to secure himself against defeat, but cannot make certain of defeating the enemy.

Hence the saying: One may *know* how to conquer without being able to do it.

Security against defeat implies defensive tactics; ability to defeat the enemy means taking the offensive.

Standing on the defensive indicates insufficient strength; attacking a super-abundance of strength.

The general who is skilled in defence hides in the most secret recesses of the earth; he who is skilled in attack flashes forth from the topmost heights of heaven. Thus on the one hand we have ability to protect ourselves; on the other, a victory that is complete.

To see victory only when it is within the ken of the common herd is not the acme of excellence. Neither is it the acme of excellence if you fight and conquer and the whole Empire says, 'Well done!'

To lift an autumn hair is no sign of great strength; to see sun and moon is no sign of sharp sight; to hear the noise of thunder is no sign of a quick ear. What the ancients called a clever fighter is one who not only wins, but excels in winning with ease.

Hence his victories bring him neither reputation for wisdom nor credit for courage. He wins his battles by making no mistakes. Making no mistakes is what establishes the certainty of victory for it means conquering an enemy that is already defeated.

Hence the skilful fighter puts himself into a position which makes defeat impossible, and does not miss the moment for defeating the enemy.

Thus it is that in war the victorious strategist seeks battle after the victory has been won, whereas he who is destined to defeat first fights and afterwards looks for victory.

The consummate leader cultivates the moral law, and strictly adheres to method and discipline; thus it is in his power to control success.

In respect of military method we have, firstly, Measurement; secondly, Estimation of quantity; thirdly, Calculation; fourthly, Balancing of chances; fifthly, Victory.

Measurement owes its existence to Earth; Estimation of quantity to Measurement; Calculation to Estimation of Quantity; Balancing of chances to Calculation; and Victory to Balancing of chances.

A victorious army opposed to a routed one is as a pound's weight placed in the scale against a single grain. The onrush of a conquering force is like the bursting of pent-up waters into a chasm a thousand fathoms deep. So much for tactical dispositions.

5

ENERGY. Sun Tzu said: The control of a large force is the same in principle as the control of a few men: it is merely a question of dividing up their numbers.

Fighting with a large army under your command is nowise different from fighting with a small one: it is merely a question of instituting signs and signals.

To ensure that your whole host may withstand the brunt of the enemy's attack and remain unshaken—this is effected by manœuvres direct and indirect.

That the impact of your army may be like a grindstone dashed against an egg—that is effected by the science of weak points and strong.

In all fighting the direct method may be used for joining battle, but indirect methods will be needed in order to secure victory.

Indirect tactics, efficiently applied, are inexhaustible as Heaven and Earth, unending as the flow of rivers and streams; like the sun and moon, they end but to begin anew; like the four seasons they pass but to return once more.

There are not more than five musical notes, yet the combinations of these five give rise to more melodies than can ever be heard. There are not more than five primary colours, yet in combination they produce more hues than can ever be seen. There are not more than five cardinal tastes, yet combinations of them yield more flavours than can ever be tasted.

In battle there are not more than two methods of attack—the direct and indirect; yet these two in combination give rise to an endless series of manœuvres. The direct and indirect lead on to each other in turn. It is like moving in a circle—you never come to an end. Who can exhaust the possibilities of their combination?

The onset of troops is like the rush of a torrent which will even roll stones along its course.

The quality of decision is like the well-timed swoop of a falcon which enables it to strike and destroy its victim.

Therefore the good fighter will be terrible in his onset, and prompt in his decision.

Energy may be likened to the bending of a cross-bow; decision, to the releasing of the trigger.

Amid the turmoil and tumult of battle there may be seeming disorder, and yet no real disorder at all; amid confusion and chaos your array may be without head or tail, yet it will be proof against defeat.

Simulated disorder postulates perfect discipline; simulated fear postulates courage; simulated weakness postulates strength.

Hiding order beneath the cloak of disorder is simply a question of subdivision; concealing courage under a show of timidity presupposes a fund of latent energy; masking strength with weakness is to be effected by tactical dispositions.

Thus one who is skilful at keeping the enemy on the move maintains deceitful appearances according to which the enemy will act.

By holding out baits he keeps him on the march; then with a body of picked men he lies in wait for him.

The clever combatant looks to the effect of combined energy, and does not require too much from individuals. Hence his ability to pick out the right men and to utilize combined energy.

When he utilizes combined energy, his fighting men become as it were like unto rolling logs or stones. For it is the nature of a log or stone to remain motionless on level ground, and to move when on a slope; if four cornered, to come to a standstill, but if round-shaped to go rolling down.

Thus the energy developed by good fighting men is as the momentum of a round stone rolled down a mountain thousands of feet in height. So much on the subject of energy.

6

WEAK POINTS AND STRONG. Sun Tzu said: Whoever is first in the field and awaits the coming of the enemy will be fresh for the fight; whoever is second in the field and has to hasten to the battle will arrive exhausted.

Therefore the clever combatant imposes his will on the enemy, but does not allow the enemy's will to be imposed on him.

By holding out advantages to him he can cause the enemy to approach of his own accord; or by inflicting damage he can make it impossible for the enemy to draw near.

If the enemy is taking his ease he can harass him; if well supplied he can starve him out; if quietly encamped he can force him to move.

Appear at points which the enemy must hasten to defend; march swiftly to places where you are not expected.

An army may march great distances without distress if it marches through country where the enemy is not.

You can be sure of succeeding in your attacks if you attack places which are not defended. You can insure the safety of your defence if you hold only positions that cannot be attacked.

Hence the general is skilful in attack whose opponent does not know what to defend; and he is skilful in defence whose opponent does not know what to attack.

O divine art of subtlety and secrecy! Through you we learn to be invisible, through you inaudible; and hence hold the enemy's fate in our hands.

You may advance and be absolutely irresistible, if you make for the enemy's weak points; you may retire and be safe from pursuit if your movements are more rapid than those of the enemy.

If we wish to fight, the enemy can be forced to an engagement even though he be sheltered behind a high rampart and a deep ditch. All we need to do is to attack some other place which he will be obliged to relieve.

If we do not wish to fight, we can prevent the enemy from engaging us even though the lines of our encampment be merely traced on the ground. All we need to do is to throw something odd and unaccountable in his way.

By discovering the enemy's dispositions and remaining invisible ourselves we can keep our forces concentrated while the enemy must be divided.

We can form a single united body, while the enemy must split up into fractions. Hence there will be a whole pitted against separate parts of a whole, which means that we shall be many to the enemy's few.

And if we are thus able to attack an inferior force with a superior one our opponents will be in dire straits.

The spot where we intend to fight must not be made known; for then the enemy will have to prepare against a possible attack at several different points; and his forces being thus distributed in many directions, the numbers we shall have to face at any given point will be proportionately few.

For should the enemy strengthen his van, he will weaken his rear; should he strengthen his rear, he will weaken his van; should he strengthen his left, he will weaken his right; should he strengthen his right, he will weaken his left. If he sends reinforcements everywhere, he will be everywhere weak.

Numerical weakness comes from having to prepare against possible attacks; numerical strength from compelling our adversary to make these preparations against us.

Knowing the place and time of the coming battle we may concentrate from great distances in order to fight.

But if neither time nor place be known, then the left wing will be impotent to succour the right, the right equally impotent to succour the left, the van unable to relieve the rear, or the rear to support the van. How much more so if the furthest portions of the army are anything under a hundred *li* apart, and even the nearest are separated by several *li*.

Though according to my estimate the soldiers of Yüeh exceed our own in number, that shall advantage them nothing in the matter of victory. I say then that victory can be achieved.

Though the enemy be stronger in numbers, we may prevent him from fighting. Scheme so as to discover his plans and the likelihood of their success.

Rouse him and learn the principle of his activity or inactivity. Force him to reveal himself so as to find out his vulnerable spots.

Carefully compare the opposing army with our own so that you may know where strength is super-abundant and where it is deficient.

In making tactical dispositions the highest pitch you can attain is to conceal them; conceal your dispositions and you will be safe from the prying of the subtlest of spies, from the machinations of the wisest brains.

How victory may be produced for them out of the enemy's own tactics—that is what the multitude cannot comprehend.

All men can see these tactics whereby I conquer, but what none can see is the strategy out of which victory is evolved.

Do not repeat the tactics which have gained you one victory, but let your methods be regulated by the infinite variety of circumstances.

Military tactics are like unto water; for water in its natural course runs away from high places and hastens downwards. So in war the way to avoid what is strong is to strike what is weak.

Water shapes its course according to the ground over which it flows; the soldier works out his victory in relation to the foe whom he is facing.

Therefore just as water retains no constant shape, so in warfare there are no constant conditions.

He who can modify his tactics in relation to his opponent and thereby succeed in winning, may be called a heaven-born captain.

The five elements [1] are not always equally prominent; the four seasons make way for each other in turn. There are short days and long; the moon has its periods of waning and waxing.

7

MANŒUVRING. Sun Tzu said: In war the general receives his commands from the sovereign.

Having collected an army and concentrated his forces, he must blend and harmonize the different elements thereof before pitching his camp.

After that comes the tactical manœuvring, than which there is

---
[1] Water, fire, wood, metal, earth.

nothing more difficult. The difficulty of tactical manœuvring consists in turning the devious into the direct, and misfortune into gain.

Thus, to take a long circuitous route, after enticing the enemy out of the way, and though starting after him, to contrive to reach the goal before him, shows knowledge of the artifice of *deviation*.

Manœuvring with an army is advantageous; with an undisciplined multitude most dangerous.

If you set a fully equipped army in march in order to snatch an advantage, the chances are that you will be too late. On the other hand to detach a flying column for the purpose involves the sacrifice of its baggage and stores.

Thus if you order your men to roll up their buff-coats, and make forced marches without halting day or night, covering double the usual distance at a stretch, doing a hundred *li* in order to wrest an advantage, the leaders of your three divisions will fall into the hands of the enemy.

The stronger men will be in front, the jaded ones will fall behind, and on this plan only one-tenth of your army will reach its destination.

If you march fifty *li* in order to outmanœuvre the enemy, you will lose the leader of your first division, and only half your force will reach its goal.

If you march thirty *li* with the same object, two-thirds of your army will arrive.

We may take it then that an army without its baggage train is lost; without provisions it is lost; without bases of supply it is lost.

We cannot enter into alliances until we are acquainted with the designs of our neighbours.

We are not fit to lead an army on the march unless we are familiar with the face of the country—its mountains and forests, its pitfalls.

We shall be unable to turn natural advantages to account unless we make use of local guides.

In war practise dissimulation and you will succeed. Move only if there is a real advantage to be gained.

Whether to concentrate or to divide your troops must be decided by circumstances.

Let your rapidity be that of the wind, your compactness that of the forest. In raiding and plundering be like fire, in immovability like a mountain.

Let your plans be dark and impenetrable as night and when you move fall like a thunderbolt.

When you plunder a countryside let the spoil be divided amongst your men; when you capture new territory cut it up into allotments for the benefit of the soldiery.

Ponder and deliberate before you make a move.

He will conquer who has learnt the artifice of deviation. Such is the art of manœuvring.

The *Book of Army Management* says: On the field of battle the spoken word does not carry far enough: hence the institution of gongs and drums. Nor can ordinary objects be seen clearly enough: hence the institution of banners and flags.

Gongs and drums, banners and flags are means whereby the ears and eyes of the host may be focused on one particular point.

The host thus forming a single united body, it is impossible either for the brave to advance alone or for the cowardly to retreat alone. This is the art of handling large masses of men.

In night-fighting, then, make much use of signal fires and drums, and in fighting by day of flags and banners, as a means of influencing the ears and eyes of your army.

A whole army may be robbed of its spirit; a commander-in-chief may be robbed of his presence of mind.

Now a soldier's spirit is keenest in the morning; but noonday it has begun to flag; and in the evening his mind is bent only on returning to camp.

A clever general, therefore, avoids an army when its spirit is keen, but attacks it when it is sluggish and inclined to return. This is the art of studying moods.

Disciplined and calm, to await the appearance of disorder and hubbub amongst the enemy—this is the art of retaining self-possession.

To be near the goal while the enemy is still far from it, to wait at ease while the enemy is toiling and struggling, to be well fed while the enemy is famished—this is the art of husbanding one's strength.

To refrain from intercepting an enemy whose banners are in perfect order, to refrain from attacking an army drawn up in calm and confident array—this is the art of studying circumstances.

It is a military axiom not to advance uphill against the enemy, nor to oppose him when he comes downhill.

Do not pursue an enemy who simulates flight; do not attack soldiers whose temper is keen.

Do not swallow a bait offered by the enemy. Do not interfere with an army that is returning home.

When you surround an army leave an outlet free. Do not press a desperate foe too hard.

Such is the art of warfare.

8

VARIATION OF TACTICS. Sun Tzu said: In war the general receives his commands from the sovereign, collects his army, and concentrates his forces.

When in difficult country do not encamp. In country where highroads intersect join hands with your allies. Do not linger in dangerously isolated positions. In hemmed-in situations you must resort to stratagem. In a desperate position you must fight.

There are roads which must not be followed, armies which must not be attacked, towns which must not be besieged, positions which must not be contested, commands of the sovereign which must not be obeyed.

The general who thoroughly understands the advantages that accompany variation of tactics knows how to handle his troops.

The general who does not understand these may be well acquainted with the configuration of the country, yet he will not be able to turn his knowledge to practical account.

So the student of war who is unversed in the art of varying his plans, even though he be acquainted with the Five Advantages will fail to make the best use of his men.

Hence in the wise leader's plans considerations of advantage will be blended together. If our expectation of advantage be tempered in this way we may succeed in accomplishing the essential part of our schemes.

If, on the other hand, in the midst of difficulties we are always ready to seize an advantage, we may extricate ourselves from misfortune.

Reduce the hostile chiefs by inflicting damage on them; make trouble for them, and keep them constantly engaged; hold out specious allurements, and make them rush to any given point.

The art of war teaches us to rely not on the likelihood of the enemy's not coming, but on our own readiness to receive him; not on the chance of his not attacking, but rather on the fact that we have made our position unassailable.

There are five dangerous faults which may affect a general: (1) Recklessness, which leads to destruction; (2) cowardice, which leads to capture; (3) a hasty temper that can be provoked by insults; (4) a delicacy of honour that is sensitive to shame; (5) over-solicitude for his men, which exposes him to worry and trouble. These are the five besetting sins of a general, ruinous to the conduct of war.

When an army is overthrown and its leader slain the cause will surely be found among the five dangerous faults. Let them be a subject of meditation.

9

THE ARMY ON THE MARCH. Sun Tzu said: We now come to the question of encamping the army, and observing signs of the enemy. Pass quickly over mountains, and keep in the neighbourhood of valleys.

Camp in high places. Do not climb heights in order to fight. So much for mountain warfare.

After crossing a river you should get far away from it.

When an invading force crosses a river in its onward march, do not advance to meet it in mid-stream. It will be best to let the army get across and then deliver your attack.

If you are anxious to fight, you should not go to meet the invader near a river which he has to cross.

Moor your craft higher up than the enemy and facing the sun. Do not move upstream to meet the enemy. So much for river warfare.

In crossing salt-marshes your sole concern should be to get over them quickly without any delay.

If forced to fight in a salt-marsh, you should have the water and grass near you, and get your back to a clump of trees. So much for operations in salt-marshes.

In dry, level country take up an easily accessible position with rising ground to your right and on your rear, so that the danger may be in front and safety lie behind. So much for campaigning in flat country.

These are the four useful branches of military knowledge which enabled the Yellow Emperor to vanquish four several sovereigns.

All armies prefer high ground to low, and sunny places to dark. If you are careful of your men, and camp on hard ground, the army will be free from disease of every kind, and this will spell victory.

When you come to a hill or a bank, occupy the sunny side, with the slope on your right rear. Thus you will at once act for the benefit of your soldiers and utilize the natural advantages of the ground.

When, in consequence of heavy rains up-country, a river which you wish to ford is swollen and flecked with foam, you must wait until it subsides. Country in which there are precipitous cliffs with torrents running between, deep natural hollows, confined places, tangled thickets, quagmires and crevasses, should be left with all possible speed and not approached.

While we keep away from such places we should get the enemy to approach them; while we face them we should let the enemy have them on his rear.

If in the neighbourhood of your camp there should be any hilly country, ponds surrounded by aquatic grass, hollow basins filled with reeds, or woods with thick undergrowth, they must be carefully routed out and searched; for these are places where men in ambush or insidious spies are likely to be lurking.

When the enemy is close at hand and remains quiet, he is relying on the natural strength of his position.

When he keeps aloof and tries to provoke a battle, he is anxious for the other side to advance.

If his place of encampment is easy of access, he is tendering a bait.

Movement amongst the trees of a forest shows that the enemy is advancing. The appearance of a number of screens in the midst of thick grass means that the enemy wants to make us suspicious.

The rising of birds in their flight is the sign of an ambuscade. Startled beasts indicate that a sudden attack is coming.

When there is dust rising in a high column it is the sign of chariots advancing; when the dust is low, but spread over a wide area, it betokens the approach of infantry. When it branches out in different directions it shows that parties have been sent out to collect firewood. A few clouds of dust moving to and fro signify that the army is encamping.

Humble words and increased preparations are signs that the enemy is about to advance. Violent language and driving forward as if to the attack are signs that he will retreat.

When the light chariots come out and take up a position on the wings it is a sign that the enemy is forming for battle.

Peace proposals unaccompanied by a sworn covenant indicate a plot.

When there is much running about it means that the critical moment has come.

When some are seen advancing and some retreating it is a lure.

When soldiers stand leaning on their spears they are faint from want of food.

If those who are sent to draw water begin by drinking themselves, the army is suffering from thirst.

If the enemy sees an advantage to be gained and makes no effort to secure it the soldiers are exhausted.

If birds gather on any spot it is unoccupied. Clamour by night betokens nervousness.

If there is disturbance in the camp the general's authority is weak. If the banners and flags are shifted about sedition is afoot. If the officers are angry it means that the men are weary.

When an army feeds its horses with grain and kills its cattle for food, and when the men do not hang their cooking pots over the camp-fires, showing that they will not return to their tents, you may know that they are determined to fight to the death.

The sight of men whispering together in small knots and speaking in subdued tones points to dissatisfaction amongst the rank and file.

Too frequent rewards signify that the enemy is at the end of his resources; too many punishments betray a condition of dire distress.

To begin by bluster, but afterwards to take fright at the enemy's numbers, shows supreme lack of intelligence.

When convoys are sent with compliments in their mouths it is a sign that the enemy wishes for a truce.

If the enemy's troops march up angrily and remain facing ours for a long time without either joining battle or taking themselves off again, the situation is one that demands great vigilance and circumspection.

If our troops are no more in number than the enemy, that is amply sufficient; it means that no direct attack can be made. What we can do is simply to concentrate all our available strength, keep a close watch on the enemy, and obtain reinforcements.

He who exercises no forethought but makes light of his opponents is sure to be captured by them.

If soldiers are punished before they have grown attached to you they will not prove submissive; and, unless submissive, they will be practically useless. If, when the soldiers have become attached to you, punishments are not enforced they will still be useless.

Therefore soldiers must be treated in the first instance with humanity, but kept under control by means of iron discipline. This is a certain road to victory.

If in training soldiers' commands are habitually enforced the army will be well disciplined.

If a general shows confidence in his men but always insists on his orders being obeyed, the gain will be mutual.

10

TERRAIN. Sun Tzu said: We may distinguish six kinds of terrain, to wit: (1) Accessible ground; (2) entangling ground; (3) temporizing ground; (4) narrow passes; (5) precipitous heights; (6) positions at a great distance from the enemy.

Ground which can be freely traversed by both sides is called *accessible*. With regard to ground of this nature, be before the enemy in occupying the raised and sunny spots, and carefully guard your line of supplies. Then you will be able to fight with advantage.

Ground which can be abandoned but is hard to reoccupy is called *entangling*. From a position of this sort, if the enemy is unprepared, you may sally forth and defeat him. But if the enemy is prepared for your coming, and you fail to defeat him, then return being impossible, disaster will ensue.

When the position is such that neither side will gain by making the first move it is called *temporizing* ground. In a position of this sort even though the enemy should offer us an attractive bait, it will be advisable not to stir forth, but rather to retreat, thus enticing the enemy in his turn; then, when part of his army has come out, we may deliver our attack with advantage.

With regard to *narrow passes*, if you can occupy them first, let them be strongly garrisoned and await the advent of the enemy. Should the enemy forestall you in occupying a pass, do not go after him if the pass is fully garrisoned, but only if it is weakly garrisoned.

With regard to *precipitous heights*, you should occupy the raised and sunny spots, and there wait for him to come up. If the enemy has occupied them before you, do not follow him, but retreat to entice him away.

If you are situated at a great distance from the enemy, and the strength of the two armies is equal, it is not easy to provoke a battle, and fighting will be to your disadvantage.

These six are the principles connected with Earth. The general who holds a responsible post must study them.

Now an army is exposed to six several calamities, not arising from natural causes but from faults for which the general is responsible. These are: (1) flight; (2) insubordination; (3) collapse; (4) ruin; (5) disorganization; (6) rout.

Other conditions being equal, if one force is hurled against another ten times its size the result will be the *flight* of the former.

When the common soldiers are too strong and their officers too weak, the result is *insubordination*. When the officers are too strong and the common soldiers too weak the result is *collapse*.

When the higher officers are angry and insubordinate, and on meeting the enemy give battle on their own account from a feeling of resentment, before the commander-in-chief can tell whether or not he is in a position to fight, the result is *ruin*.

When the general is weak and without authority; when his orders are not clear and distinct; when there are no fixed duties assigned to officers and men, and the ranks are formed in a slovenly haphazard manner, the result is utter *disorganization*.

When a general, unable to estimate the enemy's strength, allows an inferior force to engage a larger one, or hurls a weak detachment against a powerful one, and neglects to place picked soldiers in the front rank, the result must be a *rout*.

These are six ways of courting defeat, which must be carefully noted by the general who has attained a responsible post.

The natural formation of the country is the soldier's best ally; but a power of estimating the adversary, of controlling the forces of victory, and of shrewdly calculating difficulties, dangers, and distances, constitutes the test of a great general.

He who knows these things, and in fighting puts his knowledge into practice, will win his battles. He who knows them not, nor practises them, will surely be defeated.

If fighting is sure to result in victory, then you must fight, even though the ruler forbid it; if fighting will not result in victory, then you must not fight even at the ruler's bidding.

The general who advances without coveting fame and retreats without fearing disgrace, whose only thought is to protect his country and do good service for his sovereign, is the jewel of the kingdom.

Regard your soldiers as your children, and they will follow you into the deepest valleys; look on them as your own beloved sons, and they will stand by you even unto death.

If, however, you are indulgent, but unable to make your authority felt; kind-hearted but unable to enforce your commands; and incapable, moreover, of quelling disorder, then your soldiers must be likened to spoilt children; they are useless for any practical purpose.

If we know that our own men are in a condition to attack, but are unaware that the enemy is not open to attack, we have gone only half-way towards victory.

If we know that the enemy is open to attack, but are unaware that our own men are not in a condition to attack, we have gone only half-way towards victory.

If we know that the enemy is open to attack, and also know that our own men are in a condition to attack, but are unaware that the nature of the ground makes fighting impracticable, we have gone only half-way towards victory.

Hence the experienced soldier, once in motion, is never bewildered; once he has broken camp, he is never at a loss.

Hence the saying: If you know the enemy and know yourself, your victory will not stand in doubt; if you know Heaven and know Earth you may make your victory complete.

II

THE NINE SITUATIONS. Sun Tzu said: The art of war recognizes nine varieties of ground: (1) Dispersive ground; (2) facile ground; (3) contentious ground; (4) open ground; (5) ground of intersecting highways; (6) serious ground; (7) difficult ground; (8) hemmed-in ground; (9) desperate ground.

When a chieftain is fighting in his own territory it is dispersive ground.

When he has penetrated into hostile territory, but to no great distance, it is facile ground.

Ground, the possession of which imports great advantage to either side, is contentious ground.

Ground on which each side has liberty of movement is open ground.

Ground which forms the key to three contiguous states, so that he who occupies it first has most of the Empire at his command, is ground of intersecting highways.

When an army has penetrated into the heart of a hostile country, leaving a number of fortified cities in his rear, it is serious ground.

Mountain forests, rugged steeps, marshes, and fens—all country that is hard to traverse, this is difficult ground.

Ground which is reached through narrow gorges, and from which we can retire only by tortuous paths, so that a small number of the enemy would suffice to crush a large body of our men, this is hemmed-in ground.

Ground on which we can only be saved from destruction by fighting without delay is desperate ground.

On dispersive ground, therefore, fight not. On facile ground, halt not. On contentious ground, attack not.

On open ground do not try to block the enemy's way. On ground of intersecting highways join hands with your allies.

On serious ground gather in plunder.

On hemmed-in ground resort to stratagem. On desperate ground fight.

Those who were called skilful leaders of old knew how to drive a wedge between the enemy's front and rear; to prevent co-operation between his large and small divisions; to hinder the good troops from rescuing the bad, the officers from rallying their men.

When the enemy's men were scattered they prevented them from concentrating; even when their forces were united they managed to keep them in disorder.

When it was to their advantage they made a forward move; when otherwise they stopped still.

When asked how to cope with a great host of the enemy in orderly array and on the point of marching to the attack, I should say: 'Begin by seizing something which your opponent holds dear; then he will be amenable to your will.'

Rapidity is the essence of war; take advantage of the enemy's unreadiness, make your way by unexpected routes, and attack unguarded spots.

The following are the principles to be observed by an invading force: the further you penetrate into a country, the greater will be the solidarity of your troops, and thus the defenders will not prevail against you.

Make forays in fertile country in order to supply your army with food.

Carefully study the well-being of your men and do not overtax them. Concentrate your energy and hoard your strength. Keep your army continually on the move and devise unfathomable plans.

Throw your soldiers into positions whence there is no escape, and they will prefer death to flight. Officers and men alike will put forth their uttermost strength.

Soldiers when in desperate straits lose the sense of fear. If there is no place of refuge they will stand firm. If they are in the heart of a hostile country they will show a stubborn front. If there is no help for it they will fight hard.

Thus, without waiting to be marshalled, the soldiers will be constantly on the *qui vive*; without waiting to be asked they will do your will; without restrictions they will be faithful; without giving orders they can be trusted.

Prohibit the taking of omens, and do away with superstitious doubts. Then, until death comes, no calamity need be feared.

If our soldiers are not overburdened with money it is not because they have a distaste for riches; if their lives are not unduly long it is not because they are disinclined to longevity.

On the day they are ordered out to battle your soldiers may weep, those sitting up bedewing their garments, and those lying down letting the tears run down their cheeks. But let them once be brought to bay and they will display the courage of a Chu or a Kuei.

The skilful tactician may be likened to the *shuai-jan*. Now the shuai-jan is a snake that is found in the Ch'ang mountains. Strike at its head and you will be attacked by its tail; strike at its tail and you will be attacked by its head; strike at its middle and you will be attacked by head and tail both.

Asked if an army can be made to imitate the shuai-jan, I should answer Yes. For the men of Wu and the men of Yüeh are enemies; yet if they are crossing a river in the same boat and are caught by a storm they will come to each other's assistance just as the left hand helps the right.

Hence it is not enough to put one's trust in the tethering of horses, and the burying of chariot wheels in the ground.

The principle on which to manage an army is to set up one standard of courage which all must reach.

How to make the best of both strong and weak—that is a question involving the proper use of ground.

Thus the skilful general conducts his army just as though he were leading a single man, willy-nilly, by the hand.

It is the business of a general to be quiet and thus ensure secrecy; upright and just, and thus maintain order.

He must be able to mystify his officers and men by false reports and appearances, and thus keep them in total ignorance.

By altering his arrangements and changing his plans he keeps the enemy without definite knowledge. By shifting his camp and taking circuitous routes he prevents the enemy from anticipating his purpose.

At the critical moment the leader of an army acts like one who has climbed up a height and then kicks away the ladder behind him. He carries his men deep into hostile territory before he shows his hand.

He burns his boats and breaks his cooking pots; like a shepherd driving a flock of sheep, he drives his men this way and that, and none knows whither he is going.

To muster his host and bring it into danger: this may be termed the business of the general.

The different measures suited to the nine varieties of ground; the expediency of aggressive or defensive tactics; and the fundamental laws of human nature, are things that must most certainly be studied.

When invading hostile territory the general principle is that penetrating deeply brings cohesion; penetrating but a short way means dispersion.

When you leave your own country behind and take your army across neighbouring territory you find yourself on critical ground. When there are means of communication on all four sides the ground is one of intersecting highways.

When you penetrate deeply into a country it is serious ground. When you penetrate but a little way it is facile ground.

When you have the enemy's strongholds on your rear and narrow passes in front it is hemmed-in ground. When there is no place of refuge at all it is desperate ground.

Therefore, on dispersive ground, I would inspire my men with unity of purpose. On facile ground I would see that there is close connection between all parts of my army.

On contentious ground I would hurry up my rear.

On open ground I would keep a vigilant eye on my defences. On ground of intersecting highways I would consolidate my alliances.

On serious ground I would try to ensure a continuous stream of supplies. On difficult ground I would keep pushing on along the road.

On hemmed-in ground I would block any way of retreat. On desperate ground I would proclaim to my soldiers the hopelessness of saving their lives. For it is the soldier's disposition to offer an obstinate resistance when surrounded, to fight hard when he cannot help himself, and to obey promptly when he has fallen into danger.

We cannot enter into alliance with neighbouring princes until we are acquainted with their designs. We are not fit to lead an army on the march unless we are familiar with the face of the country—its mountains and forests, its pitfalls and precipices, its marshes and swamps. We shall be unable to turn natural advantages to account unless we make use of local guides.

To be ignorant of any one of the following four or five principles does not befit a warlike prince.

When a warlike prince attacks a powerful state his generalship shows itself in preventing the concentration of the enemy's forces. He overawes his opponents and their allies are prevented from joining against him.

Hence he does not strive to ally himself with all and sundry, nor does he foster the power of other states. He carries out his own secret designs, keeping his antagonists in awe. Thus he is able to capture their cities and overthrow their kingdoms.

Bestow rewards without regard to rule, issue orders without regard to previous arrangements and you will be able to handle a whole army.

Confront your soldiers with the deed itself; never let them know your design. When the outlook is bright bring it before their eyes; but tell them nothing when the situation is gloomy.

Place your army in deadly peril and it will survive; plunge it into desperate straits and it will come off in safety.

For it is precisely when a force has fallen into harm's way that it is capable of striking a blow for victory.

Success in warfare is gained by carefully accommodating ourselves to the enemy's purpose.

By persistently hanging on the enemy's flank we shall succeed in the long run in killing the commander-in-chief. This is called ability to accomplish a thing by sheer cunning.

On the day that you take up your command, block the frontier passes, destroy the official tallies, and stop the passage of all emissaries.

Be stern in the council chamber so that you may control the situation.

If the enemy leaves a door open you must rush in.

Forestall your opponent by seizing what he holds dear, and subtly contrive to time his arrival on the ground.

Walk in the path defined by rule, and accommodate yourself to the enemy until you can fight a decisive battle.

At first then, exhibit the coyness of a maiden, until the enemy gives you an opening; afterwards emulate the rapidity of a running hare, and it will be too late for the enemy to oppose you.

12

THE ATTACK BY FIRE. Sun Tzu said: There are five ways of attacking with fire. The first is to burn soldiers in their camp; the second is to burn stores; the third is to burn baggage-trains; the fourth is to burn arsenals and magazines; the fifth is to hurl dropping fire amongst the enemy.

In order to carry out an attack with fire we must have means available; the material for raising fire should always be kept in readiness.

There is a proper season for making attacks with fire, and special days for starting a conflagration.

The proper season is when the weather is very dry; the special days are those when the moon is in the constellations of the Sieve, the Wall, the Wing, or the Crossbar; for these are all days of rising wind.

In attacking with fire one should be prepared to meet five possible developments:

(1) When fire breaks out inside the enemy's camp, respond at once with an attack from without.

(2) If there is an outbreak of fire, but the enemy's soldiers remain quiet, bide your time and do not attack.

(3) When the force of the flames has reached its height, follow it up with an attack, if that be practicable; if not stay where you are.

(4) If it is possible to make an assault with fire from without, do not wait for it to break out within, but deliver your attack at the favourable moment.

(5) When you start a fire, be to windward of it. Do not attack from the leeward.

A wind that rises in the daytime lasts long, but a night breeze soon fails.

In every army the five developments connected with fire must be known, the movements of the stars calculated, and watch kept for the proper days.

Hence those who use fire as an aid to the attack show intelligence; those who use water as an aid to the attack gain an accession of strength.

By means of water an enemy may be intercepted but not robbed of all his belongings.

Unhappy is the fate of one who tries to win his battles and succeed in his attacks without cultivating the spirit of enterprise; for the result is waste of time and general stagnation.

Hence the saying: The enlightened ruler lays his plans well ahead; the good general cultivates his resources.

Move not unless you see an advantage; use not your troops unless there is something to be gained; fight not unless the position is critical.

No ruler should put troops into the field merely to gratify his own spleen; no general should fight a battle simply out of pique.

If it is to your advantage to make a forward move, make a forward move; if not stay where you are.

Anger may in time change to gladness; vexation may be succeeded by content.

But a kingdom that has once been destroyed can never come again into being; nor can the dead ever be brought back to life.

Hence the enlightened ruler is heedful, and the good general full of caution. This is the way to keep a country at peace and an army intact.

13

THE USE OF SPIES. Sun Tzu said: Raising a host of a hundred thousand men and marching them great distances entails heavy loss on the people and a drain on the resources of the state. The daily expenditure will amount to a thousand ounces of silver. There will be commotion at home and abroad, and men will drop down exhausted on the highways. As many as seven hundred thousand families will be impeded in their labour.

Hostile armies may face each other for years, striving for victory which is decided in a single day. This being so, to remain in ignorance of the enemy's condition simply because one grudges the outlay of a hundred ounces of silver in honours and emoluments, is the height of inhumanity.

One who acts thus is no leader of men, no present help to his sovereign, no master of victory.

Thus what enables the wise sovereign and the good general to strike and conquer, and achieve things beyond the reach of ordinary men, is *foreknowledge*.

Now this foreknowledge cannot be elicited from spirits; it cannot be obtained inductively from experience, nor by any deductive calculation.

Knowledge of the enemy's dispositions can only be obtained from other men.

Hence the use of spies, of whom there are five classes: (1) Local spies; (2) inward spies; (3) converted spies; (4) doomed spies; (5) surviving spies.

When these five kinds of spy are all at work none can discover the secret system. This is called 'divine manipulation of the threads'. It is the sovereign's most precious faculty.

Having *local spies* means employing the services of the inhabitants of a district.

Having *inward spies*, making use of officials of the enemy.

Having *converted spies*, getting hold of the enemy's spies and using them for our own purposes.

Having *doomed spies*, doing certain things openly for purposes of deception, and allowing our own spies to know of them and report them to the enemy.

Surviving spies, finally, are those who bring back news from the enemy's camp.

Hence it is that with none in the whole army are more intimate relations to be maintained than with spies. None should be more liberally rewarded. In no other business should greater secrecy be preserved. Spies cannot be usefully employed without certain intuitive sagacity.

They cannot be properly managed without benevolence and straightforwardness.

Without subtle ingenuity of mind one cannot make certain of the truth of their reports. Be subtle! and use your spies for every kind of business.

If a secret piece of news is divulged by a spy before the time is ripe he must be put to death together with the man to whom the secret was told.

Whether the object be to crush an army, to storm a city, or to assassinate an individual, it is always necessary to begin by finding out the names of the attendants, the aides-de-camp, the door-keepers, and sentries of the general in command. Our spies must be commissioned to ascertain these.

The enemy's spies who have come to spy on us must be sought out, tempted with bribes, led away, and comfortably housed. Thus they will become converted spies and available for our service.

It is through the information brought by the converted spy that we are able to acquire and employ local and inward spies.

It is owing to his information again that we can cause the doomed spy to carry false tidings to the enemy.

Lastly, it is by his information that the surviving spy can be used on appointed occasions.

The end and aim of spying in all its five varieties is knowledge of the enemy; and this knowledge can only be derived, in the first instance, from the converted spy. Hence it is essential that the converted spy be treated with the utmost liberality.

Of old the rise of the Yin dynasty was due to I Chih who had served under the Hsia. Likewise the rise of the Chou dynasty was due to Lü Ya who had served under the Yin.

Hence it is only the enlightened ruler and the wise general who will use the highest intelligence of the army for purposes of spying, and thereby they achieve great results. Spies are a most important element in war because on them depends an army's ability to move.

Made in United States
Orlando, FL
01 December 2023